SPOTLIGHT ON WEATHER AND NATURAL DISASTERS

DROUGHTS AND HEAT WAVES

THERESE M. SHEA

PowerKiDS press

NEW YORK

Published in 2019 by The Rosen Publishing Group, Inc.
29 East 21st Street, New York, NY 10010

Editor: Elizabeth Krajnik
Cover Design: Michael Flynn
Interior Layout: Reann Nye

Photo Credits: Cover Daimond Shutter/Shutterstock.com; p. 5 idiz/Shutterstock.com; p. 6 Tom Wang/Shutterstock.com; p. 7 Cvijovic Zarko/Shutterstock.com; p. 9 (top) Courtesy of the NOAA; p. 9 (bottom) Maridav/Shutterstock.com; p. 10 trekandshoot/Shutterstock.com; p. 11 Barcroft Media/Getty Images; p. 13 Tony Campbell/Shutterstock.com; p. 15 Courtesy of the Library of Congress; p. 16 MANJUNATH KIRAN/AFP/Getty Images; p. 17 symbiot/Shutterstock.com; p. 18 Justin Sullivan/ Getty Images News/Getty Images; p. 19 AJCespedes/Shutterstock.com; p. 21 Courtesy of NASA; p. 22 Pacific Press/LightRocket/Getty Images.

Library of Congress Cataloging-in-Publication Data

Names: Shea, Therese, author.
Title: Droughts and heat waves / Therese M. Shea.
Description: New York : PowerKids Press, 2019. | Series: Spotlight on weather and natural disasters | Includes index.
Identifiers: LCCN 2018001045| ISBN 9781508168911 (library bound) | ISBN 9781508168935 (pbk.) | ISBN 9781508168942 (6 pack)
Subjects: LCSH: Droughts--Juvenile literature. | Heat waves (Meteorology)--Juvenile literature.
Classification: LCC QC929.25 .S54 2019 | DDC 551.57/73--dc23
LC record available at https://lccn.loc.gov/2018001045

Manufactured in the United States of America

CPSIA Compliance Information: Batch #CS18PK For further information contact Rosen Publishing, New York, New York at 1-800-237-9932.

CONTENTS

CHAPTER ONE

FEEL THE HEAT

Imagine the sun is beating down on you. You're sweating—and you're not even moving. All you can see around you is dry, brown grass and dying plants. If this unusually hot weather has been going on for a while, it's probably a heat wave.

Heat waves can cause more problems than just ruining the grass around your home. They can kill crops, animals, and even people. Heat waves can also cause droughts, which are long periods of very dry weather. During a drought, there isn't enough rain to grow crops successfully. Water supplies may dry up as well.

Droughts and heat waves don't always occur at the same time. However, in recent years in the United States, they've been happening together more often. In this book, you'll learn more about heat waves and droughts and the **damage** each can cause.

A drought can make a hot day even hotter because the sun heats the air and land instead of evaporating water, or changing water from a liquid to a gas. A heat wave can make dry conditions drier.

CHAPTER TWO

STUDYING HEAT WAVES

A heat wave isn't just a few days of hot weather, and different organizations have their own meanings for the word "drought." The World Meteorological Organization defines a heat wave as five or more days in a row during which the temperature is higher than the average temperature for that time by 9°Fahrenheit (5°Celsius) or more. The U.S. National Weather Service

Weather systems don't change as quickly in the summer, so a heat wave may last for days or even weeks. The high-pressure system acts like a cap on a bottle: it traps hot air over an area.

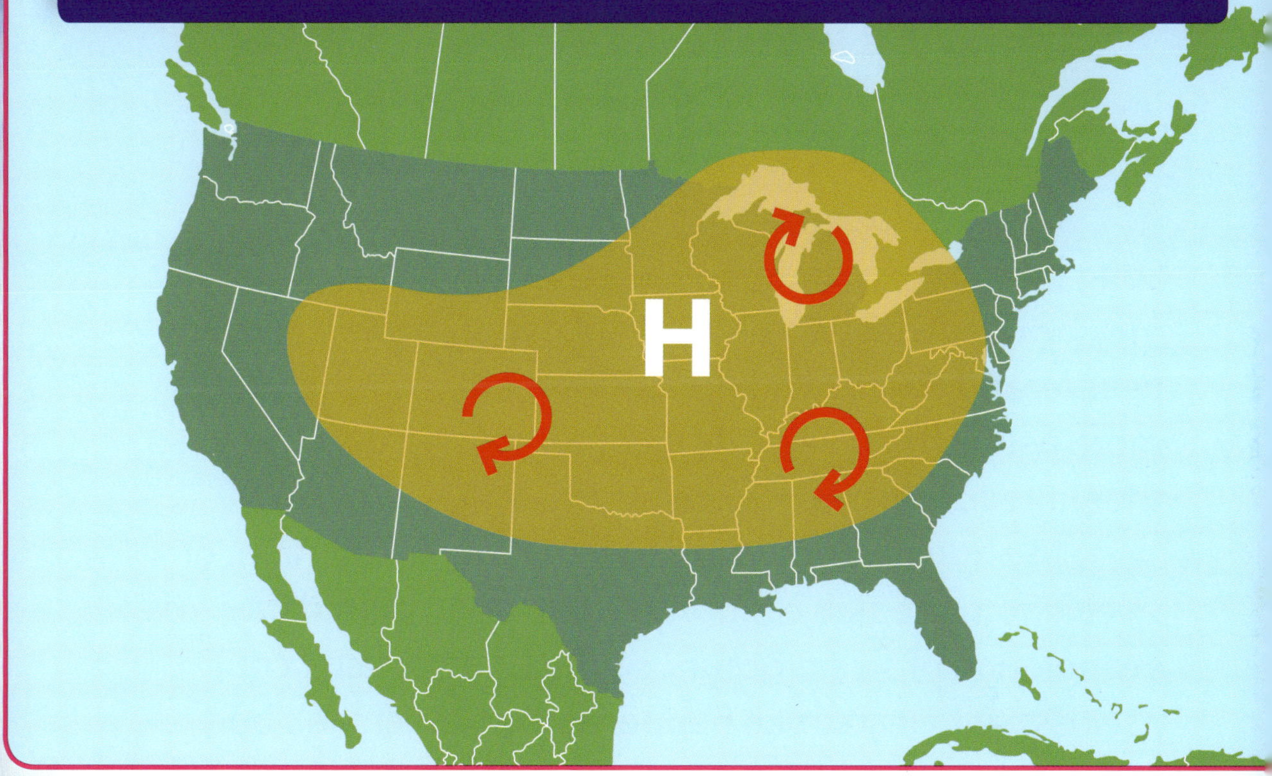

says a heat wave is "abnormally and uncomfortably hot and unusually **humid** weather" over two days or more.

Why do heat waves happen? Heat waves occur when a high-pressure system forms in the atmosphere, or the gases that surround Earth, and remains over an area. A high-pressure system forces air downward. The air can't rise and grows hotter. Without rising air, **precipitation** can't form.

CHAPTER THREE

CONSIDERING THE CONSEQUENCES

The consequences of heat waves can be serious. Long periods of high temperatures and high humidity may be deadly, especially for very young people, older people, and people with health issues. Heat waves kill more people than all other weather-related **disasters** in the United States.

The human body cools itself when perspiration, or sweat, from the skin evaporates. On a hot, humid day, less evaporation of sweat occurs, so the body doesn't cool itself as well. Heatstroke happens when a person's body temperature rises above 104°Fahrenheit (40°Celsius). This can harm the body's organs and even cause death.

Low-humidity heat waves are also a problem. During these heat waves, there's less moisture in the air. The dry air can result in a drought, which means there's less water and fewer crops.

The heat index is the temperature the body feels when heat and humidity combine. This chart from the U.S. National Oceanic and Atmospheric Administration (NOAA) shows how much hotter the air feels as humidity and temperature increase.

NOAA's National Weather Service

Heat Index

Temperature (°F)

Relative Humidity (%)	80	82	84	86	88	90	92	94	96	98	100	102	104	106	108	110
40	80	81	83	85	88	91	94	97	101	105	109	114	119	124	130	136
45	80	82	84	87	89	93	96	100	104	109	114	119	124	130	137	
50	81	83	85	88	91	95	99	103	108	113	118	124	131	137		
55	81	84	86	89	93	97	101	106	112	117	124	130	137			
60	82	84	88	91	95	100	105	110	116	123	129	137				
65	82	85	89	93	98	103	108	114	121	128	136					
70	83	86	90	95	100	105	112	119	126	134						
75	84	88	92	97	103	109	116	124	132							
80	84	89	94	100	106	113	121	129								
85	85	90	96	102	110	117	126	135								
90	86	91	98	105	113	122	131									
95	86	93	100	108	117	127										
100	87	95	103	112	121	132										

Likelihood of Heat Disorders with Prolonged Exposure or Strenuous Activity

Caution | Extreme Caution | Danger | Extreme Danger

CHAPTER FOUR

HORRENDOUS HEAT WAVES

In the summer of 2003, a heat wave fell upon western Europe. France in particular experienced abnormally hot weather in August. In some areas, temperatures were around 99°Fahrenheit (37°Celsius) for more than a week. At least 30,000 people throughout Europe—more than 14,000 people in France alone—died from heat-related causes.

In addition to deaths, the heat wave resulted in rockslides after it melted mountain snows, and dry leaves and plants helped spread forest fires. Low water levels caused **nuclear**

In 2015, a heat wave in India killed more than 2,500 people. This image shows a man cooking eggs using only the sun's heat during the heat wave.

power plants to shut down, so people weren't able to power fans and air-conditioning units.

Even deadlier was the Russian heat wave of 2010. About 55,000 people died as heat covered around 400,000 square miles (1,035,995 sq km) of the country. In Moscow, temperatures reached a record-breaking 101°Fahrenheit (38°Celsius). Wildfires added dangerous pollutants to **smog**. Heat and smog combined to kill nearly 700 people a day.

CHAPTER FIVE

DROUGHT DATA

A drought is a long period of time during which there is very little or no rain. Droughts can cause water shortages, crop damage, and reduced water both under the ground and in the soil near the surface. During a drought, more water evaporates from ground sources, including plants, than falls from the sky. Droughts that last for weeks or just a few months are considered short term, while long-term droughts can last for six or more months or even years.

However, not all droughts are unexpected. There are different kinds:

- Permanent drought is found in places that have historically dry climates.
- Seasonal drought happens in places with regularly dry and rainy seasons.
- Unpredictable drought is an unusual occurrence and usually lasts a short period.
- Invisible drought occurs when rain falls but doesn't supply enough water to replace the water lost through evaporation and **transpiration**.

Drought is a danger to agriculture in nearly every part of the world. You can see how a drought spoiled this corn crop.

CHAPTER SIX

DREADFUL DROUGHTS

Some droughts are major historical events. In fact, extended droughts may have caused the downfall of entire civilizations. Scientists believe Egypt's Old Kingdom (2575–2130 BC), the period during which the pyramids of Giza were built, came to an end at least partly because of drought.

In the 1930s, a severe drought hit the American Great Plains. Dust storms swept the region, giving the area the name "the Dust Bowl." According to the NOAA, at least 50 million acres (20.2 million ha) of land—much of it farmland—was affected. Crops, livestock, and thousands of people died.

In the 1980s, another drought began in the northern Great Plains. In 1988, drought and heat waves combined to make an even worse situation. Forest fires burned about 1.2 million acres (485,622 ha) of Yellowstone National Park. The drought caused around $39 billion of damage.

This famous photograph shows a woman and her children outside a camp of poor farm workers in Nipomo, California, in 1936 after fleeing the Dust Bowl.

CHAPTER SEVEN

COMBATING DROUGHTS

People have thought of creative ways to combat, or fight, droughts. One is cloud seeding. In this process, people place chemicals in the air to make drops of water in the atmosphere come together. These drops form ice crystals, which fall as precipitation. However, many scientists aren't convinced that cloud seeding actually works. They can't tell whether precipitation would have occurred without the cloud seeding.

CLOUD-SEEDING PLANE

Grapes can survive in dry regions, such as Arizona, because they depend more on soil makeup than climate.

Another way of taking action against droughts is to make their effects less harmful or severe. This is called drought mitigation. For example, droughts can quickly kill crops that need a lot of rain. However, by planting crops that don't need a lot of water, such as grapes and lentils, people can mitigate a drought's effects. Drought-resistant plants also don't need as much **irrigation**, which can draw water away from places where it's needed more.

CHAPTER EIGHT

WATER MANAGEMENT

One of the best ways to fight the effects of drought is to have enough water on hand so the lack of rainfall doesn't cause a disaster. People can build **reservoirs** to store water, but even reservoirs can dry up. It's wise to **conserve** water at all times so there's enough ready to use in times of need.

Conserving water can be as easy as turning off running water while you're brushing your teeth. Newer

DRIED-UP RESERVOIR IN CALIFORNIA

Drip irrigation systems supply water close to plants' roots. Older irrigation systems often spray and waste water.

dishwashers and toilets use less water. Irrigation systems for gardens and farms can also be made more water **efficient**.

In times of drought, communities have to make smart decisions about where and when to use water supplies. For example, water can help run power stations or irrigate farm crops. People also must consider that using some water sources can negatively affect plant and animal life.

CHAPTER NINE

HEAT WAVES AND CLIMATE CHANGE

Although heat waves have occurred throughout history, they're occurring more frequently now. **Climate change** is likely partly to blame. Scientists generally agree that the increase of greenhouse gases is causing global warming. Greenhouse gases include carbon dioxide, which is released into the air when we burn **fossil fuels** such as coal and oil.

A 2017 study published in *Nature Climate Change* stated the continued release of greenhouse gases into the atmosphere will lead to more and hotter heat waves. This means Earth will face greater dangers, including more droughts.

Humans will have to change their lives in order to survive. We may have to move to different areas, create **technologies** to meet the increased demand for power to cool homes and businesses, and change our farming and eating practices to lessen drought damage.

In 2009, NASA captured this image of fires in southeastern Australia that were partly caused by a heat wave and drought.

CHAPTER TEN

THE FUTURE OF EXTREME WEATHER

Interestingly, climate change doesn't only mean more heat waves and droughts. It can mean heavy rains, too. As Earth gets warmer, more water will evaporate into the air. This can cause more heavy rainstorms—and flooding from these storms. However, the evaporation will also dry out the soil so more water is needed for plant growth and water supplies. The likelihood of droughts will increase if rain doesn't fall.

Most scientists think that stopping the activities that have helped cause climate change may slow and reduce its effects, such as extreme weather. However, without changing our behavior and doing things such as swapping fossil fuels for cleaner energy sources, we can expect more heat waves and droughts. The future of Earth is in danger. How will you help fight climate change?

GLOSSARY

climate change (KLY-muht CHAYNJ) A change in Earth's weather caused by human activity.

conserve (kuhn-SUHRV) To keep something from harm and not waste it.

damage (DAA-mij) Loss or harm done to a person or piece of property.

disaster (dih-ZAS-tuhr) Something that happens suddenly and causes much suffering or loss to many people.

efficient (ih-FIH-shuhnt) Capable of producing desired results with little waste.

fossil fuel (FAH-suhl FYOOL) A fuel such as coal, oil, or natural gas that is formed in the earth from dead plants or animals.

humid (HYOO-muhd) Having a lot of moisture in the air.

irrigation (ihr-uh-GAY-shuhn) The act of supplying land with water using human-made means.

nuclear (NOO-klee-uhr) Of, relating to, producing, or using energy that is created when the smallest parts of matter are split apart or joined together.

precipitation (prih-sih-puh-TAY-shuhn) Water that falls to the ground as hail, mist, rain, sleet, or snow.

reservoir (REH-zuh-vwaar) A human-made lake that is used for storing water.

smog (SMAHG) Fog mixed with smoke and chemicals.

technology (tek-NAH-luh-jee) A method that uses science to solve problems and the tools used to solve those problems.

transpiration (trans-puh-RAY-shuhn) The process by which plants give off water vapor through openings in their leaves.

INDEX

PRIMARY SOURCE LIST

Page 15
"Destitute pea pickers in California. Mother of seven children. Age thirty-two. Nipomo, California." Photograph. Dorothea Lange. 1936. Now kept at the Library of Congress Prints and Photographs Division Washington, D.C.

Page 18
"Before And After: Statewide Drought Takes Toll On California's Lake Oroville Water Level." Photograph. Justin Sullivan. August 19, 2014. Courtesy of Getty Images.

Page 21
"Australia6 Subset - Aqua 250m True Color 2009/039 (02/08)." Satellite image. February 8, 2009. From National Aeronautics and Space Administration.

WEBSITES

Due to the changing nature of Internet links, PowerKids Press has developed an online list of websites related to the subject of this book. This site is updated regularly. Please use this link to access the list: www.powerkidslinks.com/swnd/heat